SILENT REVOLUTION

THE POWERFUL RISE OF ELECTRIC CARS

INTODUCTION

"Silent revolution" is a powerful metaphor that highlights the quiet, yet powerful impact that electric vehicles are having on the world. This also captures the idea that electric cars represent a paradigm shift in the way we think about transportation, and emphasizes the potential of electric vehicles to create a more sustainable and environmentally friendly future.

• • • • • • • • • •

CONTENTS

— • —

PREFACE

Electric vehicles have arrived, and they're here to stay. They're silent, they're efficient, and they're changing the way we think about transportation. The rise of electric cars represents a significant turning point in the automotive industry, one that will have far-reaching implications for our economy, our environment, and our way of life.

In this book, we'll explore the exciting and powerful rise of electric cars and examine the ways in which they are transforming the world around us. From the earliest electric vehicles to the latest cutting-edge models, we'll take a journey through the history of electric cars and explore the technologies that have made them possible.

But this book isn't just a history lesson. We'll also look ahead and examine the ways in which electric cars are shaping the future of transportation. We'll explore the challenges that lie ahead, from infrastructure to battery technology, and consider the ways in which we can overcome them.

Ultimately, this book is a celebration of the silent revolution that is taking place right now. It's a celebration of the people and companies that are driving this change, and of the technologies that are making it possible. We hope that after reading this book, you'll share our enthusiasm for electric cars, and that you'll be inspired to join the revolution yourself. So let's get started, and discover the powerful rise of electric cars.

• • • • ● • ● • • •

1

— ◦ —

Introduction

A. The History of Electric Vehicles

E lectric vehicles are not a new concept. In fact, the idea of using electric power to propel vehicles dates back to the early 19th century. However, it wasn't until the development of more advanced battery technology and electric motors in the late 19th and early 20th centuries that electric vehicles began to gain popularity.

The earliest electric vehicles were primitive by today's standards. They were slow, had limited range, and were expensive to produce. Nonetheless, they offered a cleaner and quieter alternative to the noisy, polluting gasoline-powered vehicles of the time.

One of the earliest electric vehicles was built in 1837 by a Scottish inventor named Robert Davidson. Davidson's electric carriage was powered by a galvanic cell and had a range of about 1.5

miles. Another early electric vehicle was built in 1881 by French inventor Gustave Trouvé. Trouvé's three-wheeled electric vehicle was capable of speeds up to 12 mph and had a range of about 30 miles.

In the United States, the first electric vehicle was built in 1891 by William Morrison. Morrison's vehicle was a six-passenger wagon that could reach speeds up to 14 mph and had a range of about 20 miles. Another early American electric vehicle was the Columbia Electric Phaeton, which was introduced in 1897. The Phaeton was capable of speeds up to 15 mph and had a range of about 40 miles.

Despite their early promise, electric vehicles faced a number of challenges in the early 20th century. One of the biggest obstacles was the high cost of batteries, which made electric vehicles prohibitively expensive for most people. Additionally, gasoline-powered vehicles were becoming more popular and affordable, and the development of more advanced internal combustion engines made gasoline-powered vehicles faster and more efficient.

However, electric vehicles continued to have a niche market throughout the 20th century. In the 1960s and 1970s, electric vehicles were used by postal services and other government agencies for short-range deliveries. In the 1990s, General Motors introduced the EV1, the first mass-produced electric vehicle in the modern era. Despite its popularity among early adopters, the EV1 was ultimately discontinued due to a lack of consumer demand and pressure from the oil industry.

Fast forward to the 21st century, and electric vehicles are once again on the rise. Improvements in battery technology and electric motors have made electric vehicles more affordable, faster, and capable of longer range than ever before. Tesla, a company founded in 2003 with the goal of accelerating the world's transition to sustainable energy, has become a leader in the electric vehicle industry. The company's Model S sedan, introduced in 2012, set a new standard for electric vehicles with its long range, high performance, and luxurious amenities.

Other automakers have also jumped on the electric vehicle bandwagon, with companies like Nissan, BMW, and Chevrolet introducing electric models to their lineups. In addition, electric vehicles are now being used for a variety of applications beyond passenger cars, including buses, trucks, and even airplanes.

Despite their history of ups and downs, electric vehicles have come a long way since their earliest days. Today, they offer a clean, quiet, and sustainable alternative to gasoline-powered vehicles, and represent a key part of our transition to a more sustainable future.

B. The Rise of Electric Vehicles in Recent Years

Electric vehicles have experienced a resurgence in popularity in recent years, driven by a combination of factors including advancements in technology, government incentives, and growing public awareness of the need for sustainable transportation.

One of the key factors contributing to the rise of electric vehicles is the rapid advancement of battery technology. Lithium-ion batteries, which are used in most modern electric vehicles, have seen significant improvements in energy density, allowing for longer range and faster charging times. This has made electric vehicles more practical for everyday use, and has helped to dispel concerns about range anxiety among consumers.

Government incentives have also played a role in the growth of the electric vehicle market. In many countries, governments offer tax incentives and rebates to consumers who purchase electric vehicles, making them more affordable and attractive to buyers. In addition, many cities are implementing policies to promote electric vehicles, such as the installation of public charging infrastructure and preferential parking for electric vehicles.

The growth of the electric vehicle market has also been driven by the increasing awareness of the environmental impact of traditional gasoline-powered vehicles. Electric vehicles produce zero emissions, making them a cleaner alternative to gasoline-powered vehicles that contribute to air pollution and greenhouse gas emissions.

Tesla, the electric vehicle manufacturer founded by Elon Musk, has been a major player in the growth of the electric vehicle market. The company's Model S sedan, introduced in 2012, set a new standard for electric vehicles with its long range, high performance, and luxurious amenities. Tesla's Model X SUV and Model 3 sedan have also been successful, with the Model 3 becoming the best-selling electric vehicle in the world in 2020.

Other automakers have also made significant strides in the electric vehicle market in recent years. Nissan's Leaf, introduced in 2010, was one of the first mass-market electric vehicles, and has since become one of the best-selling electric vehicles in the world. BMW's i3, introduced in 2013, was the first mass-produced car to use carbon fiber in its construction, making it lightweight and efficient.

In addition to passenger cars, electric vehicles are also being used for a variety of other applications. Electric buses are becoming increasingly common in cities around the world, with manufacturers such as Proterra and BYD leading the way. Electric trucks are also being developed, with companies such as Tesla, Rivian, and Nikola Motors all working on electric truck models.

In the aviation industry, electric aircraft are also being developed, with companies such as Eviation and Ampaire working on electric aircraft prototypes. While electric aircraft are still in the early stages of development, they have the potential to revolutionize air travel by offering a cleaner and quieter alternative to traditional aircraft.

Overall, the rise of electric vehicles in recent years has been driven by a combination of technological advancements, government incentives, and growing awareness of the need for sustainable transportation. As battery technology continues to improve and more automakers introduce electric models to their lineups, it's likely that electric vehicles will continue to grow in popularity and become a more common sight on our roads and highways.

C. The Benefits of Electric Vehicles

Electric vehicles offer a wide range of benefits over traditional gasoline-powered vehicles. From lower operating costs to reduced environmental impact, there are many reasons why more and more people are making the switch to electric vehicles.

Environmental Benefits

One of the primary benefits of electric vehicles is their reduced environmental impact. Electric vehicles produce zero emissions, which means they do not release harmful pollutants into the air.

This can help to improve air quality in urban areas, reducing the risk of respiratory illnesses and other health problems.

In addition, electric vehicles can also help to reduce greenhouse gas emissions. While the production of electricity used to charge electric vehicles does produce emissions, electric vehicles are still significantly cleaner than gasoline-powered vehicles over their lifetime. According to the Union of Concerned Scientists, driving an electric vehicle in the United States produces less greenhouse gas emissions than driving a gasoline-powered vehicle that gets 50 miles per gallon.

Lower Operating Costs

Electric vehicles can also offer significant savings in terms of operating costs. Electricity is typically much cheaper than gasoline on a per-mile basis, which means that electric vehicles can be much cheaper to operate over their lifetime. In addition, electric vehicles require less maintenance than gasoline-powered vehicles, as they have fewer moving parts and do not require regular oil changes or other routine maintenance.

Improved Performance

Electric vehicles also offer a number of performance benefits over traditional gasoline-powered vehicles. Electric motors can deliver instant torque, which means that electric vehicles can often accelerate more quickly than gasoline-powered vehicles. In addition, electric vehicles are typically quieter and smoother than gasoline-powered vehicles, offering a more comfortable driving experience.

Energy Security

Electric vehicles can also help to improve energy security by reducing dependence on foreign oil. While electricity can also be generated from imported sources, it can also be generated from domestic sources such as wind, solar, and hydro power. This can help to reduce the risk of price spikes and supply disruptions associated with imported oil.

Public Health Benefits

Finally, electric vehicles can also offer a number of public health benefits. By reducing emissions and improving air quality, electric vehicles can help to reduce the risk of respiratory illnesses, cardiovascular disease, and other health problems associated with air pollution. In addition, electric vehicles are typically much quieter than gasoline-powered vehicles, which can help to reduce noise pollution in urban areas.

Overall, electric vehicles offer a wide range of benefits over traditional gasoline-powered vehicles. From improved environmental impact to lower operating costs and improved performance, electric vehicles are an attractive option for anyone looking for a cleaner, more sustainable mode of transportation. As battery technology continues to improve and more automakers introduce electric models to their lineups, it's likely that electric vehicles will become an even more popular choice in the years to come.

• • • ● • ● ● • •

2

— · —

Types of Electric Vehicles

E lectric vehicles come in several different forms, each with its own unique characteristics and benefits. In this chapter, we will explore the different types of electric vehicles, starting with battery electric vehicles (BEVs).

A. Battery Electric Vehicles (BEVs)

Battery electric vehicles, or BEVs, are fully electric vehicles that are powered entirely by batteries. They have an electric motor that is powered by a large battery pack, which is typically located under the floor of the vehicle. The battery is charged by plugging the vehicle into an electrical outlet or a charging station.

BEVs offer several advantages over traditional gasoline-powered vehicles. For one, they produce zero emissions, which means they are much better for the environment. They also offer lower operating costs, as electricity is typically much cheaper than gasoline on a per-mile basis. Finally, BEVs offer a quieter and smoother driving experience, as electric motors are much quieter than gasoline engines and provide instant torque.

There are several examples of popular BEVs on the market today. One of the most well-known is the Tesla Model S, which offers impressive range and performance. The Model S can travel up to 373 miles on a single charge and can accelerate from 0 to 60 mph in just 2.4 seconds. Other popular BEVs include the Nissan Leaf, the Chevy Bolt, and the Hyundai Kona Electric.

BEVs are also being used in a variety of applications beyond personal transportation. For example, electric buses and trucks are becoming more common in urban areas as cities look for ways to reduce emissions and improve air quality. Electric delivery vans are also becoming more popular as companies look for ways to reduce operating costs and improve their environmental impact.

Overall, battery electric vehicles offer a compelling alternative to traditional gasoline-powered vehicles. With their lower operating costs, improved environmental impact, and impressive performance, it's no surprise that more and more consumers are making the switch to electric vehicles. As battery technology continues to improve, it's likely that BEVs will become even more popular in the years to come.

B. Hybrid Electric Vehicles (HEVs)

While battery electric vehicles (BEVs) offer a fully electric driving experience, hybrid electric vehicles (HEVs) combine electric power with a gasoline-powered engine. HEVs typically have a smaller battery pack than BEVs and rely on regenerative braking to recharge the battery.

HEVs offer several advantages over traditional gasoline-powered vehicles. They are more fuel-efficient, as the electric motor can assist the gasoline engine during acceleration and provide power to the wheels during low-speed driving. HEVs also produce fewer emissions than traditional gasoline-powered vehicles, as the electric motor can operate the vehicle at low speeds without using the gasoline engine.

There are several types of HEVs, including parallel hybrids, series hybrids, and plug-in hybrids. Parallel hybrids use both the electric motor and the gasoline engine to power the vehicle, while series hybrids rely solely on the electric motor to power the vehicle and use the gasoline engine as a generator to recharge the battery. Plug-in hybrids offer the ability to charge the battery from an external source, such as a charging station, and typically offer a longer electric-only range than other types of hybrids.

There are several examples of popular HEVs on the market today. One of the most well-known is the Toyota Prius, which was one of the first mass-produced hybrid vehicles. The Prius has a reputation for being extremely fuel-efficient and has sold millions of units worldwide. Other popular HEVs include the Honda Insight, the Ford Fusion Hybrid, and the Hyundai Ioniq Hybrid.

HEVs are also being used in a variety of applications beyond personal transportation. Hybrid buses and trucks are becoming more common as cities look for ways to reduce emissions and improve air quality. Hybrid systems are also being used in marine applications, such as hybrid ferries and ships, as well as in industrial applications such as hybrid cranes and forklifts.

Overall, hybrid electric vehicles offer a compelling alternative to traditional gasoline-powered vehicles, with their improved fuel efficiency and reduced emissions. As technology continues to improve, it's likely that HEVs will become even more popular in the years to come, especially as consumers become more concerned with the environmental impact of their vehicles.

C. Plug-in Hybrid Electric Vehicles (PHEVs)

Plug-in hybrid electric vehicles (PHEVs) are a type of hybrid electric vehicle that can be charged from an external source, such as a charging station, in addition to using the gasoline engine and regenerative braking to recharge the battery. PHEVs typically have a larger battery pack than traditional hybrids, which allows them to operate in electric-only mode for a longer range.

PHEVs offer several advantages over traditional gasoline-powered vehicles and traditional hybrid vehicles. They offer improved fuel efficiency compared to gasoline-powered vehicles and can operate in electric-only mode, which reduces emissions and operating costs. PHEVs also offer the flexibility of using both gasoline and electricity, which allows for longer trips without worrying about range anxiety.

There are several examples of popular PHEVs on the market today. One of the most well-known is the Toyota Prius Prime, which offers a longer electric-only range than the traditional Prius. The Prius Prime can travel up to 25 miles on a single charge and has a total range of over 600 miles. Other popular PHEVs include the Chevrolet Volt, the Ford Fusion Energi, and the Hyundai Sonata Plug-in Hybrid.

PHEVs are also being used in a variety of applications beyond personal transportation. PHEVs are being used in fleets, such as police cars and taxis, as well as in commercial applications, such as delivery vans and trucks. PHEVs are also being used in marine applications, such as hybrid ferries and ships, and in industrial applications such as hybrid cranes and forklifts.

Overall, plug-in hybrid electric vehicles offer a compelling alternative to traditional gasoline-powered vehicles and traditional hybrid vehicles. With their improved fuel efficiency, reduced emissions, and increased range, PHEVs are becoming more popular among consumers and businesses alike. As battery technology continues to improve, it's likely that PHEVs will become even more popular in the years to come.

D. Fuel Cell Electric Vehicles (FCEVs)

Fuel cell electric vehicles (FCEVs) are a type of electric vehicle that uses hydrogen fuel cells to generate electricity to power an electric motor. FCEVs emit only water vapor and have the potential to be a zero-emission vehicle. FCEVs offer several advantages over battery electric vehicles, including longer range and faster refueling times.

FCEVs work by combining hydrogen and oxygen in a fuel cell to generate electricity. The electricity is used to power an electric motor, which drives the vehicle. The only byproduct of this process is water vapor, making FCEVs a clean and environmentally friendly alternative to gasoline-powered vehicles.

One of the most well-known FCEVs on the market today is the Toyota Mirai. The Mirai has a range of up to 312 miles on a single tank of hydrogen and can be refueled in just a few minutes. Other examples of FCEVs include the Hyundai Nexo and the Honda Clarity Fuel Cell.

FCEVs are still relatively new to the market and face several challenges, including a lack of hydrogen refueling infrastructure and high production costs. However, as more automakers invest in FCEV technology and more hydrogen refueling stations are built, FCEVs have the potential to become a more viable alternative to gasoline-powered vehicles.

FCEVs also have the potential to be used in a variety of applications beyond personal transportation. FCEVs can be used in fleets, such as buses and delivery trucks, and in industrial applications such as forklifts and generators.

Overall, fuel cell electric vehicles offer a promising alternative to gasoline-powered vehicles and battery electric vehicles. While FCEVs face several challenges, including a lack of infrastructure and high production costs, continued investment in this technology has the potential to create a cleaner and more sustainable transportation future.

· · · ● ● · ● · ● · · ·

3

COMPONENTS OF AN ELECTRIC VEHICLE

A. Electric Motor

T he electric motor is one of the most important components of an electric vehicle (EV). It is responsible for converting electrical energy stored in the battery into mechanical
energy that drives the wheels of the vehicle. The electric motor
is a crucial component in the overall efficiency and performance
of an EV.

There are two main types of electric motors used in EVs: AC
(alternating current) motors and DC (direct current) motors.
AC motors are more commonly used in EVs due to their higher efficiency and power density. AC motors can also operate at
higher speeds and are more reliable than DC motors.

One of the most well-known examples of an AC motor used in an
EV is the Tesla Model S motor. The Model S uses a three-phase,

four-pole AC induction motor that delivers up to 615 horsepower and can propel the car from 0 to 60 mph in just 2.4 seconds.

DC motors, on the other hand, are simpler and less expensive than AC motors, making them a popular choice for smaller and less powerful EVs. DC motors are often used in low-speed electric vehicles, such as golf carts and neighborhood electric vehicles.

The efficiency and performance of an electric motor depend on several factors, including the design of the motor, the type of battery used, and the controller that manages the flow of electricity to the motor. Electric motors can also be designed to regeneratively brake, which means that they can convert the energy from braking into electricity that can be stored in the battery.

In addition to powering the wheels of the vehicle, electric motors can also be used to power other components in an EV, such as the air conditioning system, power steering, and other auxiliary systems. This is known as "electric accessory drive" and can help to reduce the load on the battery, increasing the range of the vehicle.

Overall, the electric motor is a crucial component of an EV, and advances in motor technology continue to play a key role in improving the efficiency and performance of electric vehicles. As more automakers invest in EV technology and battery technology improves, we can expect to see even more advanced electric motors in future EVs.

B. Battery

The battery is perhaps the most critical component of an electric vehicle (EV). It provides the power necessary to run the electric motor, as well as all of the other electrical systems in the vehicle. The battery also plays a significant role in determining the range and performance of an EV.

There are several types of batteries used in EVs, each with its own advantages and disadvantages. The most common types of EV batteries include:

Lithium-ion batteries: These are currently the most popular type of battery used in EVs due to their high energy density, long life, and relatively low cost. Lithium-ion batteries are commonly used in EVs such as the Tesla Model S and Model X, as well as the Nissan Leaf and Chevrolet Bolt.

Nickel-metal hydride (NiMH) batteries: NiMH batteries were one of the first types of batteries used in EVs and are still used in some hybrid electric vehicles (HEVs). They have a lower energy density than lithium-ion batteries, but they are more durable and have a longer lifespan. NiMH batteries are commonly used in the Toyota Prius and Honda Insight.

Solid-state batteries: Solid-state batteries are a relatively new technology that has the potential to improve the range and charging time of EVs. They use a solid electrolyte instead of a liquid electrolyte, which can provide higher energy density and faster charging times. Companies such as Toyota and Volkswagen are investing in solid-state battery technology.

Lead-acid batteries: Lead-acid batteries are an older technology that is still used in some low-speed electric vehicles (LSVs) such as golf carts and neighborhood electric vehicles. They are inexpensive but have a short lifespan and low energy density.

The performance and range of an EV are highly dependent on the battery capacity, which is measured in kilowatt-hours (kWh). The higher the battery capacity, the farther an EV can travel on a single charge. For example, the Tesla Model S has a battery capacity of up to 100 kWh, which allows it to travel up to 402 miles on a single charge.

Battery management systems (BMS) are used to monitor the health of the battery and optimize its performance. BMS can monitor battery temperature, charge and discharge rates, and the state of charge (SOC). They can also control the flow of electricity to the motor and other electrical systems to ensure the battery is not overcharged or over-discharged, which can damage the battery.

As EV technology continues to advance, battery technology is likely to play an increasingly important role in improving the range and performance of EVs. There is ongoing research into new battery chemistries and materials, as well as improvements in manufacturing processes and charging infrastructure, which will help to make EVs more accessible and practical for consumers.

C. Charging System

One of the primary concerns of electric vehicle (EV) owners is charging. Unlike conventional gasoline vehicles, EVs require a reliable and convenient system for charging their batteries. The charging system for an EV is an essential component that is responsible for supplying power to the battery to keep the vehicle running. In this chapter, we will explore the different types of EV charging systems, how they work, and their pros and cons.

Level 1 Charging: Level 1 charging is the slowest and most basic type of charging, and it involves plugging an EV into a standard 120-volt electrical outlet. Level 1 charging typically adds 2-5 miles of range per hour of charging, making it best suited for vehicles with lower daily driving needs. Level 1 charging is inexpensive, and it can be done at home or at most public charging stations. However, it is not practical for most EV owners, as it takes a long time to charge a depleted battery fully.

Level 2 Charging: Level 2 charging is a more powerful and faster charging option that is commonly used by EV owners. Level 2 chargers require a 240-volt electrical supply and are capable of adding up to 30 miles of range per hour of charging. Level 2 chargers can be installed at home or in public locations, such as parking garages, shopping malls, and restaurants. They are more expensive than level 1 chargers, but they are still relatively affordable and offer a more practical charging solution for most EV owners.

DC Fast Charging: DC Fast Charging is the fastest type of charging available for EVs, capable of adding up to 200 miles of range in as little as 30 minutes of charging. DC Fast Chargers use direct current (DC) to quickly charge an EV's battery. DC Fast Chargers are typically found at public charging stations along highways, making long-distance travel in an EV possible. However, DC Fast Chargers are more expensive than Level 1 and Level 2 chargers, and they are not as widely available as the other two options.

In addition to the different levels of charging, there are also various connector types used for charging EVs. The most common connector types used in North America are:

J1772: This connector is used for Level 1 and Level 2 charging and is the standard connector for most EVs in North America.

CCS (Combo): CCS (Combined Charging System) is used for DC Fast Charging and is a combination of the J1772 connector and an additional two DC pins for high-speed charging.

CHAdeMO: CHAdeMO is also used for DC Fast Charging and is primarily used by Japanese automakers, including Nissan and Mitsubishi.

Tesla Supercharger: Tesla has its own proprietary DC Fast Charging connector, which is only compatible with Tesla vehicles.

Charging an EV is a straightforward process that involves plugging the vehicle into the charging station and waiting for the battery to charge fully. Most EVs also have a built-in charging timer, which allows owners to schedule charging during off-peak hours when electricity rates are lower.

As EV technology continues to advance, the charging system is also evolving, with new technologies being developed to make charging faster and more convenient. For example, wireless charging technology is being developed, which will allow EVs to charge simply by parking over a charging pad. The future of EV charging is exciting, with new developments promising to make charging more accessible and convenient for EV owners.

D. Power Electronics

In an electric vehicle, the power electronics system is responsible for managing the flow of electrical energy between the battery, the motor, and other electrical components. Power electronics convert the direct current (DC) from the battery into alternating current (AC) to power the electric motor, and also convert AC power generated by regenerative braking back into DC for storage in the battery. The efficiency and performance of an electric vehicle's power electronics can greatly impact its overall range and performance.

Power electronics typically consist of three main components: the inverter, the converter, and the DC-DC converter.

The inverter is responsible for converting the DC power from the battery into AC power for the electric motor. The inverter uses a series of transistors to switch the DC power on and off at high frequencies, creating an AC signal that can be used by the motor. The inverter must be highly efficient to minimize power losses and maximize the vehicle's range.

The converter is responsible for converting the AC power generated by regenerative braking back into DC power for storage in the battery. The converter works in a similar manner to the inverter, using a series of transistors to switch the AC power on and off at high frequencies. The converter must also be highly efficient to maximize the amount of energy that can be recovered during regenerative braking.

The DC-DC converter is responsible for converting the high-voltage DC power from the battery into the low-voltage DC power needed to operate other electrical components in the vehicle, such as the lights, radio, and air conditioning. The DC-DC converter must also be highly efficient to minimize power losses and maximize the vehicle's range.

One of the challenges of designing power electronics for electric vehicles is managing the high temperatures generated by the components. The high current and voltage levels can generate significant amounts of heat, which can degrade the performance and reliability of the electronics. To address this issue, power electronics are often equipped with cooling systems, such as liquid cooling or air cooling.

Overall, power electronics are a critical component of electric vehicles, and advancements in this technology are essential to improving the efficiency, range, and performance of electric vehicles.

E. Thermal Management System

In an electric vehicle, the thermal management system is responsible for regulating the temperature of the battery, motor, and power electronics. Maintaining the correct operating temperature range is critical for ensuring optimal performance, range, and longevity of these components.

The thermal management system typically consists of a combination of liquid cooling and/or air-cooling systems, along with sensors and controllers to monitor and adjust the temperature as needed. The cooling system can be designed to cool the components during normal operation, and also to provide additional cooling during rapid charging or high demand driving situations.

Battery thermal management is particularly important, as lithium-ion batteries are sensitive to high temperatures and can degrade quickly if exposed to prolonged periods of heat. Overheating can also increase the risk of fire or explosion. Therefore, most electric vehicles use liquid cooling systems to maintain the temperature of the battery pack.

The cooling system circulates a coolant through channels or passages in the battery pack, which absorbs heat from the cells and carries it away to a radiator or heat exchanger. The coolant is typically a mixture of water and antifreeze, similar to the coolant used in a conventional combustion engine vehicle. Some electric vehicles also use refrigerant-based cooling systems, which are more efficient at removing heat but also more complex and expensive.

The thermal management system also plays a critical role in managing the temperature of the electric motor and power electronics. High temperatures can degrade the performance and reliability of these components and can also reduce the efficiency of the motor. Therefore, many electric vehicles use liquid cooling systems to maintain the temperature of the motor and power electronics.

In addition to cooling, the thermal management system can also provide heating to the cabin and battery pack during cold weather. Electric vehicles typically use electric resistance heaters, which convert electrical energy from the battery into heat to warm the cabin. Some electric vehicles also use heat pumps, which are more efficient and can provide both heating and cooling.

Overall, the thermal management system is a critical component of electric vehicles, and advancements in this technology are essential to improving the performance, range, and durability of electric vehicles. Improvements in cooling and heating efficiency can also help to reduce the energy consumption of electric vehicles, further improving their overall sustainability.

· • • ●• • ● • • ·

4

— · —

ENVIRONMENTAL AND ECONOMIC BENEFITS

A. Reduction in greenhouse gas emissions

One of the primary benefits of electric vehicles is their ability to significantly reduce greenhouse gas emissions compared to conventional gasoline or diesel vehicles. The transportation sector is one of the largest contributors to global greenhouse gas emissions, and electric vehicles offer a promising solution to help mitigate the effects of climate change.

Electric vehicles produce zero tailpipe emissions, meaning they do not emit any pollutants or greenhouse gases from their exhaust systems. This is in contrast to gasoline or diesel vehicles, which emit significant amounts of carbon dioxide, nitrogen oxides, particulate matter, and other pollutants.

Even when taking into account the emissions from the production of electricity used to charge electric vehicles, studies have shown that electric vehicles are still significantly cleaner than gasoline or diesel vehicles. In fact, electric vehicles can reduce greenhouse gas emissions by up to 70% compared to conventional vehicles on a life cycle basis, which includes emissions from the production of the vehicle, the production of the fuel or electricity used, and the operation and disposal of the vehicle.

The amount of emissions reduction depends on the source of electricity used to charge the vehicle. In regions with a high proportion of renewable energy sources such as wind, solar, and hydropower, electric vehicles can produce nearly zero emissions. Even in regions with a higher proportion of fossil fuel-based electricity, electric vehicles still produce significantly fewer emissions than conventional vehicles.

The reduction in greenhouse gas emissions from electric vehicles can have significant environmental and economic benefits. Electric vehicles can help to improve air quality in urban areas, where air pollution from vehicles is a significant health concern. They can also reduce dependence on fossil fuels, which are a finite resource and subject to price volatility and geopolitical instability.

In addition, electric vehicles can provide economic benefits by reducing the amount of money spent on imported oil and by creating jobs in the manufacturing, maintenance, and installation of electric vehicle infrastructure.

Overall, the environmental and economic benefits of electric vehicles are significant and offer a promising solution to some of the most pressing challenges facing the transportation sector today. By reducing greenhouse gas emissions and improving air quality, electric vehicles can help to create a cleaner, healthier, and more sustainable future.

B. Energy Security and Independence

Electric vehicles offer significant potential for improving energy security and independence by reducing dependence on foreign oil and diversifying energy sources. The transportation sector is one of the largest consumers of oil, and reducing oil consumption is crucial for reducing dependence on foreign oil and ensuring energy security.

Electric vehicles use electricity as their primary energy source, which can be generated from a variety of domestic sources, including renewable energy sources such as solar, wind, and hydropower, as well as domestic fossil fuels such as coal and natural gas. By using a mix of domestic energy sources, electric vehicles can help to reduce dependence on foreign oil and increase energy security.

In addition, electric vehicles offer the potential to shift energy demand away from peak periods, reducing the need for new power plants and transmission infrastructure. By charging during off-peak periods, such as at night when demand for electricity is lower, electric vehicles can help to balance the grid and reduce the need for additional energy infrastructure.

Furthermore, electric vehicles can also serve as a source of distributed energy storage, providing grid services such as frequency regulation and peak shaving. This can help to improve the stability and reliability of the electric grid, reducing the risk of blackouts and other disruptions.

The energy security and independence benefits of electric vehicles are particularly relevant in countries that are heavily dependent on oil imports, such as the United States. In 2019, the United States imported 9.1 million barrels of crude oil per day, making it the world's largest oil importer. By reducing dependence on foreign oil, electric vehicles can help to improve the country's energy security and reduce the risk of disruptions caused by geopolitical events or price fluctuations.

Electric vehicles also offer the potential to create new economic opportunities and jobs in the manufacturing and installation of electric vehicle infrastructure, as well as in the domestic production of energy sources such as renewable energy and domestic fossil fuels.

Overall, electric vehicles offer significant potential for improving energy security and independence by reducing dependence on foreign oil and diversifying energy sources. By using a mix of domestic energy sources and serving as a source of distributed energy storage, electric vehicles can help to create a more stable, reliable, and resilient electric grid, while also creating new economic opportunities and jobs.

C. Cost Savings over the Lifetime of the Vehicle

One of the most compelling benefits of electric vehicles is the potential for significant cost savings over the lifetime of the vehicle. While electric vehicles may have a higher upfront cost than their gasoline-powered counterparts, they offer substantial savings in operating costs, maintenance, and fuel costs over the life of the vehicle.

Operating Costs:

Electric vehicles have significantly lower operating costs than gasoline-powered vehicles. Electric vehicles have fewer moving parts than traditional gasoline-powered vehicles, resulting in lower maintenance costs. Electric vehicles do not require oil changes, spark plug replacements, or other routine maintenance associated with gasoline-powered vehicles. In addition, electric vehicles have regenerative braking, which reduces wear on brake pads and results in lower maintenance costs.

Fuel Costs:

Electric vehicles offer substantial savings on fuel costs compared to gasoline-powered vehicles. Electric vehicles can be charged using electricity from the grid, which is typically much cheaper than gasoline. The cost of electricity varies depending on location, but in many cases, the cost of electricity can be as low as one-third the cost of gasoline on a per-mile basis.

In addition, electric vehicles can be charged at home, eliminating the need to visit gas stations. This can be especially convenient for individuals who live in areas where gas stations are not easily accessible. Public charging stations are also becoming more common, making it easier for electric vehicle owners to charge their vehicles while on the go.

Maintenance Costs:

Electric vehicles have fewer moving parts than gasoline-powered vehicles, resulting in lower maintenance costs. Electric vehicles do not require oil changes, spark plug replacements, or other routine maintenance associated with gasoline-powered vehicles. In addition, electric vehicles have regenerative braking, which reduces wear on brake pads and results in lower maintenance costs.

Over the lifetime of the vehicle, electric vehicles can offer significant cost savings compared to gasoline-powered vehicles. The savings in operating costs, maintenance costs, and fuel costs can offset the higher upfront cost of electric vehicles. According to a study conducted by Consumer Reports, the lifetime ownership costs of electric vehicles are often less expensive than gasoline-powered vehicles.

Moreover, incentives and rebates offered by governments, as well as lower operating costs, have made electric vehicles more affordable for consumers. In some countries, such as Norway, electric vehicles are exempt from certain taxes and tolls, making them more attractive to buyers.

In conclusion, electric vehicles offer substantial cost savings over the lifetime of the vehicle. Lower operating costs, maintenance costs, and fuel costs can offset the higher upfront cost of electric vehicles. The potential for cost savings, combined with environmental and energy security benefits, makes electric vehicles an attractive option for many consumers.

D. Reduced Noise Pollution

One of the most significant benefits of electric vehicles (EVs) is their quiet operation, which results in a reduction in noise pollution. Traditional gasoline-powered vehicles are noisy and emit a considerable amount of noise pollution, particularly in urban areas, where traffic noise is a significant problem. In contrast, electric vehicles are incredibly quiet and emit little to no noise, making them an ideal solution for reducing noise pollution.

EVs are powered by electric motors that are incredibly quiet, and as a result, the sound they produce is minimal compared to conventional vehicles. The only sound that electric vehicles emit is the sound of the tires rolling on the road surface, which is considerably less than the sound generated by the engine and exhaust of a traditional vehicle. This reduction in noise pollution not only benefits the environment but also improves the quality of life for people living in urban areas.

The reduction in noise pollution has many advantages. Firstly, it can help reduce stress levels and improve overall health. Studies have shown that exposure to noise pollution can increase stress levels and contribute to various health problems such as heart disease, high blood pressure, and sleep disturbance. Therefore, the reduction in noise pollution due to electric vehicles can improve public health.

Secondly, reduced noise pollution can lead to a more peaceful and enjoyable environment, particularly in cities. It is estimated that over 50% of the world's population lives in urban areas, and the noise generated by traffic is a significant contributor to noise pollution in these areas. The quiet operation of electric vehicles can lead to a more pleasant and peaceful environment for urban residents.

Thirdly, reduced noise pollution can benefit wildlife. Wildlife living near urban areas are often affected by noise pollution, which can disrupt their behavior and cause stress. The reduction in noise pollution due to electric vehicles can help mitigate this problem, allowing wildlife to thrive in their natural habitats.

Lastly, the reduction in noise pollution due to electric vehicles can lead to the development of new and innovative urban spaces. For instance, public parks and outdoor spaces that were previously too noisy for use can now be utilized for recreation and social activities.

In conclusion, the reduction in noise pollution due to electric vehicles is a significant benefit that can improve public health, create a more peaceful and enjoyable environment, benefit wildlife, and lead to the development of new and innovative urban spaces. This benefit, combined with other environmental and economic benefits of electric vehicles, makes them a compelling solution for transportation in the 21st century.

•••••••••••

5

—— • ——

CHARGING INFRASTRUCTURE

A. Types of chargers

One of the main concerns for electric vehicle (EV) drivers is the availability of charging infrastructure. While home charging is the most convenient option for many EV owners, it's important to have access to public charging stations, especially for long-distance travel. The charging infrastructure has been rapidly expanding to meet the growing demand for EVs. This chapter will discuss the different types of chargers that are currently available.

Level 1 Chargers

Level 1 chargers are the most basic charging option for electric vehicles. They are designed to be plugged into a standard 120-volt household outlet and can provide up to 5 miles of range per hour of charging. Level 1 chargers are typically included with the purchase of an electric vehicle and are convenient for overnight charging at home. However, they are not recommended for regular use as they are slow and cannot fully charge the vehicle within a reasonable amount of time.

Level 2 Chargers

Level 2 chargers are more powerful than Level 1 chargers and can provide up to 25 miles of range per hour of charging. They require a 240-volt electrical outlet, which is typically found in homes with electric dryers or ranges. Level 2 chargers are the most common type of charging station found in public locations such as parking lots, shopping centers, and workplaces. They can fully charge an EV in about 4-6 hours, making them a convenient option for longer trips.

DC Fast Chargers

DC fast chargers are the fastest and most powerful charging option available for electric vehicles. They can provide up to 300 miles of range per hour of charging and can fully charge an EV in as little as 30 minutes. DC fast chargers require specialized charging equipment and are typically found at highway rest stops, gas stations, and other locations where drivers can quickly charge their vehicles during long-distance travel. While DC fast chargers are the most convenient option for long trips, they are more expensive to install and maintain than Level 1 and Level 2 chargers.

Wireless Charging

Wireless charging is an emerging technology that allows EVs to charge without the need for cables or plugs. It works by using an electromagnetic field to transfer energy from a charging pad to the vehicle's battery. While wireless charging is still in its early stages, it has the potential to make EV charging even more convenient and accessible. Currently, wireless charging is only available in a few select locations and is limited to a few models of EVs.

In conclusion, the availability of charging infrastructure is essential for the widespread adoption of electric vehicles. Level 1 and Level 2 chargers are the most common types of chargers and can be found in homes and public locations. DC fast chargers are the fastest and most powerful option, but are more expensive to install and maintain. Wireless charging is an emerging technology that has the potential to make EV charging even more convenient in the future.

B. Public Charging Stations

While electric vehicles can be charged at home, there is still a need for public charging infrastructure to support longer trips and to accommodate those who do not have access to home charging.

Public charging stations come in a variety of types and locations, from highway rest areas to shopping centers and parking garages. In this chapter, we will explore the different types of public charging stations and their features.

Types of Public Charging Stations:

Level 1 Charging: This is the slowest type of charging, typically using a standard 120-volt household outlet. Level 1 charging can take up to 20 hours to fully charge an electric vehicle, making it impractical for anything other than emergency charging.

Level 2 Charging: This is the most common type of public charging station and is typically found in public parking lots, office buildings, and shopping centers. Level 2 charging stations use a 240-volt power supply and can fully charge an electric vehicle in 4-8 hours, depending on the size of the battery.

DC Fast Charging: DC Fast Charging, also known as Level 3 charging, is the fastest type of charging available for electric vehicles. These charging stations use a 480-volt power supply and can charge an electric vehicle to 80% capacity in 30 minutes or less. DC Fast Charging stations are typically found along highways and major travel corridors.

Features of Public Charging Stations:

Payment Options: Public charging stations can be free to use, or they may require payment. Payment options can vary, with some charging stations requiring payment by credit card, while others use a smartphone app or require a membership.

Connectivity: Some charging stations offer Wi-Fi connectivity, allowing electric vehicle owners to access the internet while their vehicle charges.

Location: Public charging stations can be found in a variety of locations, including highway rest areas, parking garages, and shopping centers. The location of charging stations can be important for electric vehicle owners who need to charge while on a long trip.

Availability: While the number of public charging stations is growing, they can still be limited in some areas. Availability can be a concern for electric vehicle owners who need to charge while away from home.

Maintenance: Like any piece of infrastructure, public charging stations require maintenance. Some charging stations are maintained by the charging station owner, while others are maintained by a third-party service provider.

In conclusion, public charging stations play a critical role in the adoption of electric vehicles. While home charging is the most convenient and cost-effective option for many electric vehicle owners, public charging infrastructure is essential for longer trips and for those who do not have access to home charging. The availability and convenience of public charging stations will continue to be a key factor in the growth of the electric vehicle market.

C. Home Charging Solutions

One of the most convenient aspects of owning an electric vehicle is the ability to charge it at home. Home charging solutions can vary depending on the needs and preferences of the EV owner. In this chapter, we will discuss the different types of home charging solutions available for EV owners.

Level 1 Charging

Level 1 charging is the slowest and simplest way to charge an EV. It uses a standard household outlet (120 volts) and takes longer to charge the vehicle's battery. The average charging rate for Level 1 is about 4-5 miles of range per hour of charging. This is a good option for those who do not drive much and have plenty of time to charge their vehicle overnight.

Level 2 Charging

Level 2 charging is the most common and popular home charging solution for EV owners. It uses a 240-volt outlet, which is the same outlet used for large household appliances such as a dryer or oven. The charging rate for Level 2 is significantly faster than Level 1, averaging around 25-30 miles of range per hour of charging. Level 2 charging stations can be installed at home, making it easy for EV owners to charge their vehicle overnight.

Level 3 Charging

Level 3 charging, also known as DC fast charging, is not typically used for home charging solutions due to its high cost and power requirements. Level 3 charging stations use a 480-volt DC power source and can charge an EV battery to 80% in as little as 30 minutes. These types of charging stations are typically found in public areas such as highways, rest stops, and commercial parking lots.

Charging Equipment

EV owners can choose from a variety of charging equipment options depending on their needs and preferences. Some popular charging equipment options include:

Portable Chargers: Portable chargers are a convenient and easy way to charge an EV when on the go. These chargers can be plugged into a standard household outlet and can be used to charge the vehicle's battery.

Wall-Mounted Chargers: Wall-mounted chargers are a popular option for Level 2 charging at home. These chargers can be mounted on a garage wall or other convenient location and provide a faster charging rate than Level 1 charging.

Smart Chargers: Smart chargers are equipped with advanced features such as the ability to monitor and manage charging remotely. Some smart chargers can even be programmed to charge during off-peak hours to take advantage of lower electricity rates.

Cost Considerations

When considering a home charging solution, it's important to consider the cost of the equipment and installation. The cost of a Level 2 charging station can range from $500 to $2,000, depending on the features and power requirements. Installation costs can also vary depending on the complexity of the installation and the location of the charger.

Overall, home charging solutions are a convenient and cost-effective way for EV owners to charge their vehicles. With the right charging equipment and installation, EV owners can enjoy the benefits of electric driving without the need for frequent trips to public charging stations.

D. Smart Charging Systems

As the popularity of electric vehicles (EVs) continues to grow, so does the need for efficient and effective charging solutions. One promising solution is the use of smart charging systems. These systems enable EV owners to charge their vehicles at the most convenient and cost-effective times, while also reducing the strain on the electrical grid. In this chapter, we will explore the various types of smart charging systems and their benefits.

A smart charging system is an intelligent system that allows EVs to be charged in a controlled and optimized manner. These systems use advanced algorithms to schedule charging based on factors such as the EV owner's preferences, the availability of renewable energy sources, and the load on the electrical grid. The aim of these systems is to reduce the peak demand on the grid, which can help to avoid power outages and reduce the need for expensive upgrades to the electrical infrastructure.

There are several types of smart charging systems available on the market today. The most common types include:

Time-of-Use (TOU) Charging: This type of system charges the EV at specific times of the day when electricity is cheaper. EV owners can program their vehicles to charge during off-peak hours when demand is lower, and electricity prices are lower.

Dynamic Pricing: This system uses real-time electricity prices to determine the best time to charge the EV. The system charges the EV when prices are low and delays charging when prices are high.

Vehicle-to-Grid (V2G) Charging: V2G charging enables EVs to store electricity and sell it back to the grid during periods of high demand. This helps to reduce the strain on the grid and can provide additional income for EV owners.

Demand Response: This system enables EV owners to respond to peak demand events on the grid by reducing or delaying charging. This can help to reduce the strain on the grid and avoid power outages.

The benefits of smart charging systems are significant. They can reduce the cost of charging EVs, help to avoid power outages, and reduce the need for expensive upgrades to the electrical infrastructure. They can also increase the use of renewable energy sources, as EVs can be charged during periods when renewable energy sources are available. Additionally, they can help to reduce the carbon footprint of EVs by enabling them to charge using renewable energy sources.

Smart charging systems are also beneficial for utilities and grid operators. By reducing peak demand on the grid, these systems can help to avoid power outages and reduce the need for expensive upgrades to the electrical infrastructure. They can also increase the use of renewable energy sources, which can help to reduce the carbon footprint of the grid.

In conclusion, smart charging systems are a promising solution to the challenges of charging EVs. These systems enable EV owners to charge their vehicles at the most convenient and cost-effective times while reducing the strain on the electrical grid. With the continued growth of the EV market, smart charging systems are likely to become an increasingly important component of the EV ecosystem.

• • • • ● • ● • • •

6

— • —

CHALLENGES AND LIMITATIONS

A. Range Anxiety

One of the most significant challenges facing electric vehicle adoption is range anxiety. Range anxiety refers to the fear that an electric vehicle's battery will run out of charge before the driver reaches their destination, leaving them stranded.

Range anxiety is primarily caused by two factors: limited driving range and inadequate charging infrastructure. Many electric vehicles have a driving range of around 100-300 miles on a single charge, depending on the make and model. While this range is sufficient for most daily commutes and errands, it may not be enough for long road trips or traveling to remote areas with limited charging options.

Additionally, the availability and accessibility of charging stations can vary widely depending on the location. Some areas have a robust network of public charging stations, while others have very few. This can make it challenging for electric vehicle drivers to plan longer trips and navigate unfamiliar areas.

To address these issues, automakers and charging infrastructure providers are working to increase the range of electric vehicles and expand the charging network. Some automakers are developing electric vehicles with longer ranges, such as the Tesla Model S, which has a range of up to 402 miles on a single charge. Additionally, charging infrastructure providers are working to install more charging stations in high-traffic areas and along popular travel routes.

In addition to range anxiety, other challenges and limitations facing electric vehicles include the following:

Cost: Electric vehicles can be more expensive upfront than their gasoline counterparts, although they may be less expensive to operate over the vehicle's lifetime.

Charging time: Electric vehicles take longer to charge than refueling a gasoline vehicle. While most home charging stations can fully charge a vehicle overnight, public fast-charging stations can take around 30 minutes to charge a vehicle to 80% capacity.

Cold weather performance: Cold temperatures can negatively impact an electric vehicle's range and performance, particularly when the battery is not preheated before use.

Battery degradation: Over time, an electric vehicle's battery may lose capacity, reducing the vehicle's range and requiring costly battery replacements.

Despite these challenges, electric vehicles are becoming increasingly popular as more people recognize the benefits they offer. As the technology continues to evolve and the charging infrastructure expands, it is likely that range anxiety and other limitations will become less of an issue.

B. Battery Charging Time

One of the most significant limitations of electric vehicles is the time it takes to charge their batteries. While gasoline-powered cars can be refueled in a matter of minutes, charging an electric vehicle can take much longer. This is one of the main reasons why many people still choose not to purchase an electric vehicle, as the prospect of waiting for hours to charge their car is not feasible for many.

There are several factors that affect the charging time of an electric vehicle. The first is the type of charger being used. Level 1 chargers, which are typically used in home charging applications, can take up to 20 hours to fully charge a depleted battery. Level 2 chargers, which are more commonly found in public charging stations and some home installations, can charge an electric vehicle in 4-8 hours, depending on the size of the battery and the charging rate.

Fast chargers, also known as DC fast chargers or Level 3 chargers, can charge an electric vehicle much more quickly. These chargers use direct current (DC) power to quickly charge the battery, and can fully charge some electric vehicles in as little as 30 minutes. However, fast chargers are not as widely available as Level 2 chargers, and are typically only found at public charging stations or along major highways.

Another factor that affects charging time is the size of the battery. Larger batteries will take longer to charge than smaller ones, as they require more energy to be replenished. The charging rate of the charger itself is also a factor. Some chargers have a higher charging rate than others, which means they can charge the battery more quickly.

Despite the limitations of battery charging time, there are several solutions being developed to address this challenge. One solution is the development of higher-capacity batteries that can hold more energy and take less time to charge. Another solution is the development of ultra-fast charging technologies, which can charge an electric vehicle in a matter of minutes rather than hours.

One of the most promising developments in this area is the use of solid-state batteries, which use a solid electrolyte instead of a liquid electrolyte. Solid-state batteries are expected to have a higher energy density than current lithium-ion batteries, which means they can hold more energy in a smaller space. They are also expected to have a faster charging time, as they can be charged at higher rates without overheating.

In conclusion, while battery charging time is a significant limitation of electric vehicles, there are several solutions being developed to address this challenge. As the technology continues to advance, it is likely that charging times will continue to decrease, making electric vehicles more convenient and practical for everyday use.

C. Limited Availability of Charging Infrastructure

One of the major challenges faced by the widespread adoption of electric vehicles is the limited availability of charging infrastructure. While there has been a steady increase in the number of public charging stations, there are still not enough to meet the growing demand for electric vehicles. This lack of charging infrastructure is a significant barrier to the adoption of electric vehicles, particularly for those who do not have access to home charging solutions.

The availability of charging infrastructure varies widely depending on the location. Urban areas typically have a higher density of charging stations, while rural areas and smaller towns have fewer options. This can make it difficult for electric vehicle owners to plan long-distance trips or travel to areas where charging stations are scarce.

Another challenge with the limited availability of charging infrastructure is the wait time to use a charging station. With the increasing popularity of electric vehicles, it is not uncommon to see long lines of electric vehicles waiting to charge at public charging stations. This wait time can be frustrating for electric vehicle owners and can discourage others from making the switch to electric vehicles.

One solution to the limited availability of charging infrastructure is the installation of fast charging stations. Fast charging stations can provide a full charge in as little as 30 minutes, making it more convenient for electric vehicle owners to charge their vehicles on the go. Additionally, some companies are working on developing wireless charging technology, which could eliminate the need for physical charging stations altogether.

Another solution is the implementation of smart charging systems. These systems can help manage the demand for charging stations and ensure that electric vehicles are charged during off-peak hours when electricity is less expensive. This can help alleviate some of the strain on the grid during peak hours and make the most efficient use of the available charging infrastructure.

Governments and private companies are recognizing the need to invest in charging infrastructure to support the transition to electric vehicles. Many countries have introduced policies and incentives to encourage the installation of charging stations, such as tax breaks and subsidies for businesses and individuals who install charging stations.

In conclusion, the limited availability of charging infrastructure remains a significant challenge for the widespread adoption of electric vehicles. However, there are solutions being developed and implemented to address this challenge, such as fast charging stations and smart charging systems. With continued investment and innovation in charging infrastructure, electric vehicles can become a more practical and accessible transportation option for people around the world.

D. Higher upfront cost compared to conventional vehicles

While electric vehicles have many benefits, there is no denying that they are more expensive than their conventional counterparts. The higher cost is often cited as a significant barrier to their adoption, but it is important to consider the lifetime cost of owning an electric vehicle.

The upfront cost of an electric vehicle is higher due to several factors, including the cost of the battery, which is the most expensive component of the vehicle. The price of batteries has been declining over the years, but they still account for a significant portion of the overall cost of an electric vehicle. However, as battery technology continues to improve, we can expect the cost of electric vehicles to come down.

In addition to the battery, electric vehicles have other components that are more expensive than those in conventional vehicles. For example, electric motors are more expensive to manufacture than internal combustion engines. Also, the power electronics and thermal management systems needed in electric vehicles are more complex and costly.

However, it is essential to consider the lifetime cost of owning an electric vehicle when evaluating the higher upfront cost.

While electric vehicles are more expensive initially, they have lower operating costs over time due to the lower cost of electricity compared to gasoline and the fewer maintenance requirements. Electric vehicles have fewer moving parts, and therefore, require less maintenance than gasoline-powered vehicles. According to a study by Consumer Reports, electric vehicle owners can save up to 50% on maintenance and repair costs over five years compared to gasoline-powered cars.

Moreover, there are incentives and tax credits available for electric vehicle buyers that can offset the higher upfront cost. Federal and state governments offer incentives such as tax credits, rebates, and grants to encourage the adoption of electric vehicles. For example, in the United States, the federal government provides a tax credit of up to $7,500 for electric vehicle buyers, depending on the battery size and the vehicle's manufacturer.

Another factor that can make the upfront cost of an electric vehicle more manageable is the lower cost of ownership. According to an analysis by Consumer Reports, over a five-year period, electric vehicles can have a lower total cost of ownership than gasoline-powered cars, even when factoring in the higher purchase price.

In conclusion, the higher upfront cost of electric vehicles compared to conventional vehicles is undoubtedly a challenge. However, it is important to consider the lifetime cost of owning an electric vehicle, which includes the lower operating and maintenance costs. Additionally, incentives and tax credits are available to help make the purchase of an electric vehicle more affordable. As battery technology improves and economies of scale come into play, we can expect the cost of electric vehicles to continue to decline, making them more accessible to a broader range of consumers.

• • • • • • • • • •

7

GOVERNMENT POLICIES AND INCENTIVES

A. Federal policies and incentives

E lectric vehicles have become an important part of the transportation industry, and governments around the world have taken notice. To encourage the adoption of electric vehicles, many governments have implemented policies and incentives to promote the purchase and use of electric vehicles. In this chapter, we will discuss the policies and incentives that the federal government of the United States has implemented to encourage the adoption of electric vehicles.

Federal Tax Credits

The federal government of the United States has implemented several tax credits to encourage the adoption of electric vehicles. The most well-known tax credit is the federal tax credit for electric vehicles. This tax credit provides up to $7,500 for the purchase of a new electric vehicle. The amount of the credit depends on the battery size of the vehicle, with larger battery sizes receiving a larger credit. However, the tax credit is limited to the first 200,000 electric vehicles sold by each manufacturer. Once a manufacturer reaches this limit, the tax credit begins to phase out.

Corporate Average Fuel Economy (CAFE) Standards

The Corporate Average Fuel Economy (CAFE) standards were first introduced in 1975 to improve the average fuel economy of cars and light trucks sold in the United States. The standards are set by the National Highway Traffic Safety Administration (NHTSA) and are designed to reduce the country's dependence on foreign oil and reduce greenhouse gas emissions.

In 2012, the CAFE standards were updated to include electric vehicles. The standards require manufacturers to meet a certain average fuel economy for their fleet of vehicles. To meet these standards, manufacturers must sell a certain percentage of electric vehicles or other zero-emission vehicles. Failure to meet these standards can result in hefty fines for the manufacturer.

Federal Grants

The federal government of the United States also offers grants to support the development and deployment of electric vehicles.

The Department of Energy (DOE) offers several grant programs to fund research and development of electric vehicles and related technologies.

One of the most well-known grant programs is the Advanced Technology Vehicles Manufacturing (ATVM) loan program. This program provides loans to manufacturers to support the production of advanced technology vehicles, including electric vehicles. The program has provided over $8 billion in loans to support the production of electric vehicles and related technologies.

Other Incentives

In addition to tax credits, CAFE standards, and grants, the federal government of the United States offers other incentives to promote the adoption of electric vehicles. These incentives include:

High-Occupancy Vehicle (HOV) Lane Access: In some states, electric vehicles are allowed to use HOV lanes regardless of the number of passengers in the vehicle.

Charging Infrastructure Funding: The federal government provides funding to support the installation of electric vehicle charging infrastructure.

Research and Development Funding: The federal government provides funding for research and development of electric vehicle technologies.

Conclusion:

The federal government of the United States has implemented several policies and incentives to encourage the adoption of electric vehicles. These policies and incentives include tax credits, CAFE standards, federal grants, and other incentives. While these policies and incentives have been successful in increasing the adoption of electric vehicles, there is still much work to be done to ensure that electric vehicles become the primary mode of transportation in the future.

B. State and Local Policies and Incentives

In addition to federal policies and incentives, many states and local governments have implemented their own programs to promote the adoption of electric vehicles. These policies and incentives vary widely across different regions and can include financial incentives, infrastructure development, and regulatory measures.

One of the most common types of state and local incentives for electric vehicles is financial incentives, which can take the form of tax credits, rebates, or grants. For example, California offers a $2,500 rebate for battery-electric vehicles and plug-in hybrid electric vehicles, and Colorado offers a $5,000 tax credit for electric vehicles. Other states, such as New York and New Jersey, offer rebates based on the vehicle's battery size and range.

Some states have also implemented infrastructure development programs to encourage the installation of charging stations. For example, Oregon has the "Electric Byways" program, which is a network of public charging stations along designated scenic routes throughout the state. Similarly, Colorado's "Charge Ahead Colorado" program provides funding to install public charging stations in designated areas.

In addition to financial incentives and infrastructure development, some states and local governments have implemented regulatory measures to promote the adoption of electric vehicles. California, for example, has set a goal of having 5 million zero-emission vehicles on the road by 2030 and has implemented a Zero Emission Vehicle (ZEV) program that requires automakers to produce a certain number of electric vehicles each year or purchase credits from other automakers that do.

Other states have implemented programs to promote the use of electric vehicles in public transportation. For example, Massachusetts has the "Clean Cities Coalition" program, which provides funding to local governments and businesses to purchase electric vehicles for use in public transportation and fleet operations.

Overall, state and local policies and incentives can play a crucial role in promoting the adoption of electric vehicles. These programs can help overcome some of the challenges and limitations associated with electric vehicles, such as the limited availability of charging infrastructure and the higher upfront cost compared to conventional vehicles. As electric vehicle technology continues to improve and become more accessible, it is likely that state and local governments will continue to play an important role in promoting the adoption of this technology.

C. International policies and initiatives

The transition to electric vehicles is not just limited to one country or region. Countries all over the world are taking steps towards promoting the use of electric vehicles to reduce emissions and achieve their climate goals. International policies and initiatives play a critical role in driving the adoption of electric vehicles worldwide.

One of the most well-known international initiatives to promote the use of electric vehicles is the Paris Agreement, which was adopted in 2015 by 196 countries. The agreement aims to limit global warming to below 2 degrees Celsius above pre-industrial levels by reducing greenhouse gas emissions. Electric vehicles are seen as an important tool in achieving these goals, and many countries have set targets for phasing out the sale of new gasoline and diesel vehicles.

Several countries have also implemented policies and incentives to promote the use of electric vehicles. In Norway, for example, electric vehicles are exempt from all non-recurring vehicle fees, including purchase taxes, and enjoy reduced annual road tax rates. Electric vehicle owners in Norway also have access to free toll roads and public charging infrastructure.

China, the world's largest market for electric vehicles, has implemented a range of policies to promote the adoption of electric vehicles. These policies include tax exemptions, subsidies for vehicle purchases, and investments in charging infrastructure. The Chinese government has also set a target for electric vehicles to make up 50% of all new car sales by 2035.

In the European Union, the Clean Vehicles Directive aims to promote the use of clean vehicles in public procurement. Under the directive, public authorities are required to purchase a minimum number of clean vehicles, including electric vehicles, for their fleets. The European Union has also set emissions standards for new cars, which will require car manufacturers to reduce the average emissions of their fleets.

Other international initiatives include the Zero Emission Vehicle Alliance, a global partnership of governments committed to promoting the adoption of electric vehicles, and the International Energy Agency's Electric Vehicle Initiative, which aims to accelerate the deployment of electric vehicles worldwide.

International policies and initiatives are crucial in promoting the adoption of electric vehicles worldwide. By creating a supportive environment for the adoption of electric vehicles, governments and organizations can accelerate the transition to a low-carbon transportation system and reduce greenhouse gas emissions.

•••••••••••

8

— ◦ —

THE FUTURE OF ELECTRIC VEHICLES

A. Technological advancements

As the demand for electric vehicles continues to grow, there are several technological advancements that are expected to shape the future of this industry. One of the most significant advancements is the development of new and more efficient batteries.

Currently, most electric vehicles use lithium-ion batteries, which have limitations in terms of energy density and range. However, there are several new battery technologies that are being developed, including solid-state batteries, lithium-sulfur batteries, and zinc-air batteries. These batteries are expected to provide greater energy density and longer range, making electric vehicles more practical and appealing to consumers.

Another important technological advancement is the development of wireless charging systems. Currently, most electric vehicles require a physical connection to a charging station, but wireless charging technology could make charging more convenient and efficient. With this technology, charging pads would be installed on roads and in parking lots, allowing electric vehicles to charge while parked or driving.

In addition to advancements in batteries and charging technology, there are also efforts to improve the overall efficiency of electric vehicles. One area of focus is regenerative braking, which allows vehicles to recover energy that would otherwise be lost during braking. This energy can be used to charge the vehicle's battery, increasing its range and reducing the need for external charging.

Another area of focus is the development of lightweight materials, such as carbon fiber and aluminum, which can reduce the weight of electric vehicles and improve their efficiency. This could lead to longer range and improved performance.

Advancements in self-driving technology are also expected to play a role in the future of electric vehicles. With autonomous driving, electric vehicles could be used more efficiently, reducing the need for personal vehicle ownership and making electric vehicles more accessible to a wider range of consumers.

Overall, the future of electric vehicles looks promising, with advancements in technology and infrastructure making these vehicles more practical and appealing to consumers. As governments and industries continue to prioritize sustainability and reducing greenhouse gas emissions, the demand for electric vehicles is likely to continue to grow, driving further innovation in this exciting field.

B. Increased consumer adoption

As more and more people become aware of the benefits of electric vehicles, consumer adoption is increasing steadily. In fact, electric vehicle sales have been growing at a rapid pace in recent years, and many experts predict that this trend will continue.

One major factor driving the increased adoption of electric vehicles is the growing concern about climate change and the desire to reduce carbon emissions. As more people become aware of the environmental benefits of electric vehicles, they are more likely to consider purchasing one for themselves. Additionally, as more electric vehicle models become available on the market, consumers have a wider range of options to choose from, which can make it easier for them to find a vehicle that meets their needs.

Another factor driving the increased adoption of electric vehicles is the decreasing cost of these vehicles. In the past, electric vehicles were often seen as too expensive for many consumers to afford, but as technology has improved and production has increased, the cost of these vehicles has come down. Additionally, many governments offer incentives and rebates for consumers who purchase electric vehicles, which can help to further reduce the cost.

Advancements in technology are also helping to increase consumer adoption of electric vehicles. For example, the range of electric vehicles is increasing, which means that drivers can travel further on a single charge. Additionally, charging times are decreasing, which makes it more convenient for drivers to charge their vehicles on the go.

Furthermore, many consumers are attracted to the advanced features and technology that electric vehicles offer. For example, many electric vehicles come equipped with advanced safety features and entertainment systems, which can make driving a more enjoyable experience. Additionally, electric vehicles are often equipped with smart features, such as the ability to connect to the internet or to sync with a smartphone, which can make it easier for drivers to stay connected and access information while on the road.

Overall, the future of electric vehicles looks bright. As more consumers become aware of the benefits of these vehicles and as technology continues to improve, it is likely that we will see even greater adoption of electric vehicles in the years to come. This could have a significant impact on our environment, our economy, and our way of life.

C. Development of new battery technologies

The development of new battery technologies is a crucial aspect of the future of electric vehicles. While the current lithium-ion batteries are efficient and reliable, they have certain limitations, such as range anxiety, charging time, and safety concerns. Thus, researchers and companies are constantly working to develop new and improved battery technologies that can overcome these limitations and make electric vehicles more practical and accessible to a wider audience.

One of the most promising battery technologies under development is solid-state batteries. Unlike the liquid or gel electrolytes used in traditional lithium-ion batteries, solid-state batteries use a solid electrolyte, which makes them safer, more efficient, and more durable. Solid-state batteries also have the potential to increase the range of electric vehicles and reduce charging times.

Another promising technology is lithium-sulfur batteries. These batteries have a higher energy density than traditional lithium-ion batteries, which means they can store more energy in a smaller and lighter package. Lithium-sulfur batteries also have the potential to be less expensive than lithium-ion batteries, which could make electric vehicles more affordable for consumers.

Other technologies being developed include zinc-air batteries, which are lightweight, low-cost, and have high energy density; flow batteries, which store energy in liquid electrolytes and can be charged quickly by replacing the electrolyte; and sodium-ion batteries, which use a low-cost and abundant material and can operate at high temperatures.

In addition to these new battery technologies, researchers are also working on improving the performance and lifespan of existing lithium-ion batteries. This includes developing new materials for the electrodes and electrolytes, as well as improving the manufacturing processes to reduce costs and increase efficiency.

The development of new battery technologies is also being supported by government and industry initiatives. In 2021, the US Department of Energy announced a goal to reduce the cost of electric vehicle batteries to $80 per kilowatt-hour by 2025, which would make electric vehicles more affordable for consumers. Additionally, companies such as Tesla, Ford, and General Motors are investing billions of dollars in research and development of new battery technologies.

Overall, the development of new battery technologies is a crucial aspect of the future of electric vehicles. These new technologies have the potential to make electric vehicles more practical, affordable, and accessible to a wider audience, which could lead to a significant reduction in greenhouse gas emissions and a shift towards a more sustainable transportation system.

D. Integration with renewable energy sources

Integration of electric vehicles with renewable energy sources is a promising trend that can help to reduce carbon emissions and increase energy efficiency. Renewable energy sources such as solar, wind, hydro, and geothermal energy can provide a clean and sustainable source of electricity to power electric vehicles.

One of the major advantages of integrating electric vehicles with renewable energy sources is that it can help to reduce the dependence on fossil fuels. By using renewable energy sources to power electric vehicles, we can reduce the need for oil imports and minimize the impact of volatile oil prices on the economy. Additionally, renewable energy sources can help to reduce carbon emissions and other pollutants associated with the burning of fossil fuels.

Solar power is one of the most promising renewable energy sources for powering electric vehicles. Solar panels can be installed on the roof of homes, businesses, and public buildings to generate electricity that can be used to power electric vehicles. Additionally, solar-powered charging stations can be installed in public places such as parking lots and rest areas to provide a clean and sustainable source of electricity for electric vehicles.

Wind power is another promising renewable energy source for powering electric vehicles. Wind turbines can be installed on land and offshore to generate electricity that can be used to power electric vehicles. Additionally, wind-powered charging stations can be installed in public places to provide a clean and sustainable source of electricity for electric vehicles.

Hydroelectric power is another promising renewable energy source for powering electric vehicles. Hydroelectric power plants generate electricity by harnessing the energy of moving water. This electricity can be used to power electric vehicles, and hydro-powered charging stations can be installed in public places to provide a clean and sustainable source of electricity for electric vehicles.

Geothermal power is another promising renewable energy source for powering electric vehicles. Geothermal power plants generate electricity by tapping into the heat energy stored in the Earth's crust. This electricity can be used to power electric vehicles, and geothermal-powered charging stations can be installed in public places to provide a clean and sustainable source of electricity for electric vehicles.

In addition to the benefits of integrating electric vehicles with renewable energy sources, there are also challenges to overcome. For example, renewable energy sources such as solar and wind can be intermittent, meaning that they may not always be available when needed. This requires the development of energy storage technologies to store excess energy for use when renewable energy sources are not available.

Overall, the integration of electric vehicles with renewable energy sources is an important trend that can help to reduce carbon emissions, increase energy efficiency, and reduce dependence on fossil fuels. As renewable energy technologies continue to improve, we can expect to see more widespread adoption of electric vehicles powered by clean and sustainable energy sources.

E. Autonomous driving and electric vehicles

Autonomous driving is a rapidly advancing technology that is set to revolutionize the automotive industry. Electric vehicles are also on the rise and are becoming more popular due to their environmental and economic benefits. The integration of these two technologies could have a profound impact on the way we travel in the future.

Autonomous driving technology has the potential to greatly enhance the driving experience of electric vehicles. One of the main benefits of autonomous driving is increased safety. According to the National Highway Traffic Safety Administration (NHTSA), 94% of crashes are caused by human error. By removing the human element from driving, autonomous vehicles could greatly reduce the number of accidents on the road.

In addition to increased safety, autonomous driving technology can also increase the efficiency of electric vehicles. Autonomous vehicles can communicate with each other and with the infrastructure around them to optimize their route and reduce energy consumption. This can lead to increased range and reduced charging time for electric vehicles.

One of the key challenges in integrating autonomous driving and electric vehicles is the development of a reliable and efficient charging infrastructure. Autonomous vehicles will need to be able to autonomously locate and access charging stations in order to maintain their charge. This will require the development of a robust network of charging stations that can be easily accessed by autonomous vehicles.

Another challenge is the development of new technologies to support the integration of autonomous driving and electric vehicles. For example, new sensors and software will need to be developed to enable autonomous vehicles to interact with charging stations and other infrastructure.

Despite these challenges, there are already several examples of autonomous electric vehicles on the market. Tesla, for example, has been working on autonomous driving technology for several years and has already implemented several features such as Autopilot and Full Self-Driving. Waymo, a subsidiary of Alphabet, is also developing autonomous electric vehicles for use in ride-sharing and delivery services.

In conclusion, the integration of autonomous driving and electric vehicles has the potential to greatly enhance the safety, efficiency, and convenience of transportation. While there are still many challenges to overcome, the rapid pace of technological development and increasing consumer demand for electric and autonomous vehicles suggest that this integration will continue to be an area of significant innovation and growth in the years to come.

• • • ● • ● • • •

— · —

CONCLUSION

A. Summary of key points

I n conclusion, electric vehicles have emerged as a promising solution to the environmental and energy challenges of our time. They have come a long way since their inception and have rapidly gained popularity in recent years.

To summarize the key points of this guide, electric vehicles are powered by electric motors and batteries, which can be charged at home or at public charging stations. They offer numerous environmental and economic benefits, such as reduced greenhouse gas emissions, energy security, and cost savings over the lifetime of the vehicle. However, they also face challenges and limitations such as range anxiety, charging time, and limited availability of charging infrastructure, as well as higher upfront costs compared to conventional vehicles.

Government policies and incentives, both at the federal and state levels, have played a crucial role in promoting the adoption of electric vehicles. International initiatives have also been launched to accelerate the transition to sustainable transportation.

The future of electric vehicles looks promising, with technological advancements such as improved batteries and charging systems, increased consumer adoption, integration with renewable energy sources, and the potential for autonomous driving. These developments hold significant potential for reducing our carbon footprint and achieving a more sustainable transportation sector.

It is clear that electric vehicles will continue to play an important role in the future of transportation, as we strive towards a more sustainable and environmentally-friendly society. As we move forward, it will be important for governments, industry leaders, and consumers to work together to overcome the challenges and realize the full potential of electric vehicles.

B. Final thoughts and reflections

Electric vehicles are rapidly becoming a viable and attractive alternative to conventional vehicles, with their numerous advantages in terms of environmental impact, energy security, and cost savings over their lifetime. The development and adoption of electric vehicles have been driven by technological advancements in batteries, motors, power electronics, and other components, as well as government policies and incentives aimed at reducing greenhouse gas emissions, improving energy security, and promoting innovation.

As electric vehicle technology continues to evolve, we can expect to see continued improvements in range, charging time, and performance, as well as greater integration with renewable energy sources and autonomous driving technologies. These developments will further increase the appeal of electric vehicles to consumers and drive even greater adoption in the years to come.

However, despite the numerous advantages of electric vehicles, there are still challenges and limitations that need to be addressed, such as range anxiety, charging infrastructure availability, and higher upfront costs compared to conventional vehicles. These challenges will need to be overcome in order to accelerate the transition to a more sustainable transportation system.

In conclusion, the future of electric vehicles is bright, with a growing market share and continuous technological advancements. While there are still obstacles to overcome, the benefits of electric vehicles make them an attractive option for consumers and policymakers alike. By continuing to invest in research and development, as well as supportive policies and incentives, we can accelerate the adoption of electric vehicles and move towards a cleaner and more sustainable transportation system.

C. Call to action for readers

As we come to the end of this comprehensive guide on electric vehicles, it's important to remember that the transition to electric mobility is not just a technological shift, but also a societal and environmental one. As consumers, policymakers, and stakeholders, we all have a role to play in accelerating the adoption of electric vehicles and driving the transition towards a more sustainable future.

If you are a consumer considering purchasing an electric vehicle, there are several steps you can take to ensure a successful transition. Firstly, research the different models available to determine which one fits your needs and budget. Secondly, consider installing a home charging station to make charging more convenient and accessible. Finally, be open to new experiences and take advantage of the many benefits that electric vehicles offer, such as reduced emissions, lower operating costs, and a quieter ride.

If you are a policymaker or stakeholder, there are also important actions you can take to support the growth of electric vehicles. One key strategy is to establish policies and incentives that promote the adoption of electric vehicles and the development of charging infrastructure. This can include subsidies, tax incentives, and regulatory measures that encourage the use of electric vehicles. Additionally, supporting research and development into new battery technologies and charging infrastructure can help drive innovation and progress in the industry.

Ultimately, the success of electric vehicles will depend on the collective efforts of all of us. By working together, we can create a cleaner, more sustainable future and build a world where electric vehicles are the norm rather than the exception. So, let's take action today and pave the way for a brighter tomorrow.

• • • ● • ● • • •